APPROACHING   THE   CHANNEL   TUNNEL

edited   by

Clive   H.   Church

European Studies Section
University of Kent at Canterbury

University Association for Contemporary European Studies

(C)   University Association for
         Contemporary European Studies

ISBN   0 906 384 23 0

UACES   Occasional Papers 3

UACES SECRETARIAT KING'S COLLEGE LONDON WC2R 2LS

2

CONTENTS

I. *INTRODUCTION*

The Channel Tunnel has been much debated in Britain, but not always positively or helpfully. It has been debated with a good deal of partisanship, and there have been few academic studies of the project. Moreover, because of partisanship, debate has concentrated on the abstract principle of whether such a fixed link is desirable or not. This has also meant that, while a good deal of attention has been given to technical and financial details, less thought has been given to how we should respond to the project when it got the go-ahead. All in all, the debate has often been more concerned with avoiding the Tunnel than with approaching it. And, for all that it is an Anglo-French project, in Britain at least, the debate has been conducted in strictly national compartments, notwithstanding the talk of the European dimension to the project.

It was in order to try and get beyond these initial barriers to understanding the significance of the Channel Tunnel that the Anglo-French Colloquium, on which this publication is based, was held. The Colloquium had its roots in the interest that the University of Kent, as not merely one of the closest Universities to the Continent but the only University in the county likely to be most affected, has in the Channel Tunnel. The latter was therefore an automatic choice for what is hoped will be the first in a series of joint colloquia between its European Studies Section and one of its exchange partners, the Institut d'Etudes Politiques in Paris. The choice was particularly apt given both the interest the IEP has in Anglo-French relations and the fact that its outgoing Director, M. Michel Gentot, had been intimately concerned with the previous Tunnel project. The two institutions were also fortunate in obtaining generous financial support, from the Cultural Section of the French Embassy in London and the Paris Office of the British Council. It was this generosity, which they are most happy to acknowledge, which enabled them to

4

meet in Canterbury from 8-10 April 1987.

Because of the interest in the Tunnel the Colloquium attracted a
wide range of other participants, both from academic institutions
in Britain (notably from Wye College which is a major centre of
research into the environmental impact of the Tunnel) and France
(especially from the Université de Lille II), and from officers
of the two local government bodies most involved in the project,
the Kent County Council and the Nord-Pas de Calais Region. The
format of the Colloquium was worked out by representatives of
the two sponsoring institutions who decided to avoid set speeches
and to concentrate on providing the conditions for intensive
discussions amongst scholars and administrators concerned with
responding to the Channel Tunnel. Hence there were three workshops each of which was presented with an introductory position
paper by a previously appointed rapporteur. Discussion was then
initiated by a discussant from the other institution. The two
then produced a draft summary of the discussions for a final
plenary session which recommended that the ideas of the
Colloquium should be published.

The academic structure of the Colloquium with its three sections
on legal and financial aspects, environmental & planning
considerations and transport & European aspects respectively,
has been maintained here on grounds of simplicity and speed. It
was felt best to publish the record of the Colloquium more or
less as it stood so no attempt has been made to update the
contributions or to eliminate some of the natural overlaps
between the sections. The book therefore contains finalized
versions of the position papers and agreed and slightly expanded
versions of the summaries of the discussions in the workshops.
Thanks are due to all those who provided or translated materials
for the Colloquium and, especially, for the book.

The idea of a wider European dimension to the project explains
in part why the papers of the Colloquium appear in the place
they do. Much has been made, on both sides of the Channel, of
the significance of the Channel Tunnel for European integration.
Pierre Mauroy has claimed that it would do more for British

insertion into Europe than the Treaty of Accession and even Mrs
Thatcher has opined that it would change the British way of life.
Other British spokesmen have talked of it ending both 8,000 years
of isolation from the continent and of it relighting the lights
over Europe whose extinguishing was lamented by Sir Edward Grey
in 1914. Even if we may be sceptical about ambitious claims for
changed perceptions and psychologies, clearly the project fits
into the British search for a form of European integration more
relevant to ordinary people than those which rely on fine
federalist phrases. Thanks to this dimension, the University
Association for Contemporary European Studies was happy to
associate itself with the Colloquium and to take responsibility
for the publication of the Colloquium papers.

The Findings of the Colloquium

The general purpose of the papers printed here, like that of the
Colloquium itself, is to go beyond rhetorical debates on 'is it
a good thing or not' or 'will the project go ahead'. They seek
to do this not by providing hard and fast answers but by thinking
realistically about some of the difficulties which will have to
be faced as we approach the construction and opening of the
Tunnel. In fact the book has three precise purposes for over-
coming some of the obstacles to a positive approach to the
problems posed by the Tunnel. It seeks, firstly, to give some
idea of the feelings of academics and administrators at the time,
and secondly to raise the questions which will have to be
answered in the not too distant future. Lastly, it seeks to
offer some suggestions for the monitoring that will also be
needed as the project develops. As a result the book is
concerned less with certainties and more with the questions which
are emerging as to how the political system will need to approach
the project as it, in turn, approaches realization. Thus it
deals less with the technological details of the project and more
with the way in which it poses problems for scholars and
administrators who seek to understand the bases of its financing,
its effects on central and local government and its likely impact
on national and European transport and economic development.

What emerged from the Colloquium is the way in which, although the project has already promoted significant Anglo-French co-operation, notably in the legal field, there are still major differences of approach and difficulties to be overcome. Sometimes these arise from different policy orientations and sometimes from a surprising lack of knowledge of the other's interests and practices. This was particularly the case in the environmental field. However, such differences were approached with a maximum of good will and each side confessed itself to be attracted by the advantages of the other's system of enquiries or site control. It thus did something to make each country more aware of the problems and proposals of the other.

Both countries also seem to experience some difficulty in adjusting to the fact that the project is to be financed, built and run by a private contractor, though both were approaching the task very positively and co-operatively. One of the problems in doing this was the different status and economic situation of the two localities most involved. This difference also accounted for the implications of the Tunnel for economic growth and European integration. Much of this was seen to depend on changing attitudes and policies in other fields and areas of government. Such questions cannot yet be confidently answered.

So it is clear that scrutiny and monitoring of the Tunnel project will be necessary as we approach its completion and opening. Similarly, administrators and academics on both sides of the Channel will need to prepare for and adapt to, the exigencies of the Tunnel. Since the Colloquium showed that there is considerable common ground and good will, the auspices for positive and successful responses in this area are good. And generally, while the Colloquium did approach some of the significant questions already raised by the debate, such as the co-ordination of the two legal systems, the treatment of investors in and competitors to the scheme, the differential effects of the two planning systems, and the intricacies of the evolving balance of the two co-operating regions, it did so in a more positive and forward-looking way than has often been the case. In this it seems to have achieved its major aim of getting

to grips with the realities of the project and it is to be hoped
that this will continue in the future.

The Evolution of the Channel Tunnel Project

However, before going on to introduce the ideas of the Colloquium
in more detail, something needs to be said about the context in
which the Colloquium and, indeed, publication took place. It
might be thought that such positive attitudes to the Tunnel were
somewhat transient and affected by the state of the Ratification
process at the time. However, while it is clear that any
discussion of the Channel Tunnel at a time when ratification and
financing were still incomplete, is bound to be highly
provisional and soon out of date, the importance of the project,
the often emotive press coverage it has attracted in Britain and
the lack of serious published analysis suggests that publication
is still likely to be of value if only as an indication of
feelings at a particular moment in the history of the project.
In the event, the discussions clearly anticipated the successful
completion of the ratification process.

The Colloquium took place at a time when there were still major
uncertainties about the evolution of the project. Although there
was more continuity between the Channel Tunnel project abandoned
in 1975 and the present scheme than is often appreciated, the
latter was clearly a surprising and unexpected proposal. It
emerged as an agreement between a Conservative cabinet in Britain
and the first Socialist government of the Fifth Republic and this
at a time when relations between the two countries were soured
by disputes over exports of lamb, apples and poultry alleged to
be suffering from Newcastle's disease. Moreover, it was a
venture which proposed to tie Britain to the Continent at a time
when public opinion was still not wholly reconciled to membership of the European Community and the government's theoretical
commitment to the Community was being called into question by
the relentless style of its pursuit of a solution to the
budgetary question. Yet although there was some delay, the
decisions to investigate the advisability of a Tunnel, to invite
tenders, and to approve the EuroTunnel concession by Treaty were
actually taken relatively rapidly. This was particularly so in

comparison to the previous project and, more importantly, given the immensely significant decision that the project should be a privately financed one carried out without direct government involvement or assistance.

Whereas in France the project was generally welcomed and considered on a technical basis, in Britain opinion was more reserved, and often apparently hostile. Large questions were raised about the political and economic implications of the Tunnel and the way in which the decision had been taken. The project was often debated as a matter of principle even though the Select Committee hearings under the hybrid bill procedure did not allow any direct questioning of the decision to proceed with a Fixed Link. The efforts of the opposition to the Tunnel, centred round the Ferry operators, seemed to have enjoyed considerable success at first. The issues raised by the opposition tended to play down the Tunnel as a facet of the Government's basic 'hands off' economic policy to concentrate on the uncertain benefits of a project which the government was seen as thrusting on the country, and the region, so as to advantage a particular commercial undertaking.

While this was happening, the Government, in order to assuage the opposition, felt it necessary to respond to suggestions from Parliamentary Committees and other interests in a variety of ways including the creation of the Joint Consultative Committee with local authorities in Kent. Similarly, the combination of hostility and the need to prepare for an unwanted eventuality led local councils and others to embark on new initiatives in order to strengthen their bargaining position and to seek to ready themselves for the eventual construction of the Tunnel. These processes were central to the Colloquium's discussions. At that stage too, although the French legislative process was almost complete, that in Britain was still grinding its way through hearings in the House of Lords. And not all that long before there had been major reconstruction of the management of the British end of EuroTunnel in order to strengthen its public relations and financial performance. Questions were still being asked about what would happen if the project failed to proceed or collapsed soon afterwards.

In the event the Bill made reasonable progress through its
British legislative scrutiny, despite the financial problems
apparently experienced by the concessionnaire. Opposition,
though not disappearing, especially locally, rather died away
and did not make much impact in the elections of May and June.
So, having been carried over into the new Parliament, the Bill
was approved and given Royal Assent towards the end of July,
with the Treaty being ratified on 29 July. This meant that the
deferred final stages of financing could go ahead in the autumn,
with major construction works following soon after. Some of
this was clearly anticipated by the Colloquium, but previous
history was still reflected in much of its discussions.

Financial and Legal Aspects

Whereas much of the debate on this side of the project had
suggested that the concessionnaire was essentially a free agent,
and that the Tunnel was a generous gift by indulgent governments,
the Workshop, both through Alain Fayard's position paper and
generally, was clear that this was not so. The project was, in
fact, enmeshed in a network of legal systems and controls. Nor
were those controls as mutually contradictory as is sometimes
supposed. They were both trying to achieve similar ends by
different means. Lawyers and politicians on both sides were also
concerned that the Tunnel should not escape their own courts and
laws. Similarly, while the financial climate now seems to be
propitious and the methods adopted for the Tunnel are being more
favourably evaluated in Europe, there are still notable financial
problems, such as the distribution of risk and the demands of
investors for protection. M. Fayard felt that the instruments
being used were sufficient but some opinion felt that the
guarantees being offered were not as solid as might be.

The questions which emerged included whether the harmonisation
of laws would prove sufficient and how the new joint legal
package would interact with domestic law. Conflicts of laws were
possible, not merely between the two countries, but between
international and local law. The problems of how differring

governmental policies on the operation of competition and market forces would be reconciled remained unclear. This may also be the case in the financial sphere where potential conflicts between security provisions and implicit cross-subsidisation on the one hand and the requirements of private financing on the other.

Monitoring of the financial evolution of the project, especially where the rights of investors and lenders are concerned, will clearly be a necessity. The concessionnaire cannot be regarded as self-sufficient and the assessment and distribution of risk remains a sensitive and debatable issue. The evaluation of the new joint legal system and the institutions set up to implement it will also require close scrutiny, as will the interaction between the various legal interests. Ratification, in other words, will not end legal and financial scrutiny of the project.

Environmental and Planning Considerations

Some of the legal provisions in the treaty were clearly relevant to the interests of the second Workshop. However, what came out most clearly in discussion was the very different systems of planning controls in the two countries and the lack of familiarity with their details which prevailed on both sides. The environmental situation on the two sides is very different and, for instance, the terrain above Dover and Folkestone does not allow for the kind of development feasible in France, where the environmental problems were fewer but technically more complex because of drainage difficulties. And while the British tend to feel that only they are concerned with environmental questions and controls on development, the evidence from France presented in the Workshop showed that this was not wholly the case. On the other hand, the discussion showed itself much more favourably inclined towards the hybrid bill procedure than is often the case, and tended to follow the arguments of Margaret Anderson's position paper on the British provisions.

The questions which emerged for future consideration centred on how the environmental concerns would be handled in the future, given that the two countries look at the project in different

ways. The French, for instance, see the TGV line as an inter-
dependent part of the project and not as something separate,
subject to normal planning rules as in Britain. Similarly, the
French were inclined to see rail as the main form of transport
to be considered, whereas in Britain roads were equally, if not
more, important. A good deal of interest was shown in the
systematic handling of both the Concessionnaire and the
construction sites provided by the 'Grand Chantier' procedure in
France, especially as it carries with it special funds. The
question of whether this, together with the greater powers of a
French region, would enable co-operation and balanced development
to take place in Kent and the Nord-Pas de Calais was clearly a
key question. The ambiguous place of Kent in the South East was
also stressed.

However, much was seen to depend on the likely demand for the
rail services offered by the Tunnel, as this would affect the
environmental impact. It was felt that a close eye would have
to be kept on the extent to which British Rail was able to
deliver on this front. Equally, the evolution of the very
different local economies and administrative structures would be
a matter of importance. For Kent it might be that the Tunnel
could be the fulcrum for better liaison with the rest. This,
and the question of economic gains from the Tunnel, which are of
major significance for environmental planning, were one of the
many points at which this Workshop overlapped with the last of
the three.

Transport and European Integration

Here much of the British debate has turned on the likely economic
impact of the project on trade and growth. However, for the
professional economist, as Roger Vickerman's position paper made
clear, this concern often rests on unfounded assumptions about
the role of transport as a factor in growth. Here too, the long
distance railway connections, linking Britain into the European
network, were likely to be more significant than the shuttle
from the European and attitudinal points of view. The discussion
in the Workshop also highlighted the different nature of the two

areas at the tunnel mouth and the broader geographic background against which the Tunnel will have to be seen.

The questions which arose were partly the question of whether the model of financing adopted will provide the price competition expected and, more particularly, the likely impact of the Tunnel on the volume, nature and location of trade and economic activity. It might be that the new axes of transport created by the Tunnel could benefit Belgium and parts of Northern France and Germany more than the coastal areas. In any case, it was clear that economic growth would not happen automatically and that attitudes and the development of specific policies would have to be considered very carefully if the two regions were to raise their profile and attract industry. The implications of the Tunnel for the completion of the Internal Market were also unclear, and this, combined with the difficulties experienced in devising an effective Common Transport Policy, implied that the impact on European integration remains to be demonstrated.

This suggests, as the Workshop saw, that monitoring of attitudes to Europeanization, especially in Britain, will be necessary. Equally, the moves made by the European institutions towards supporting and exploiting the Tunnel and the associated high speed train network will be watched with interest. But it may be that in the end the relative level of costs and benefits will be the most effective factor in determining attitudes to Europe and the Channel Tunnel, as it has been with popular responses to integration. Scrutiny of the pricing policies adopted by ferries and EuroTunnel is something else which will have to be carefully watched. But, as in so many other areas, there are still more questions than answers.

****

Although sharing concerns for the financial and economic viability of the project, the views expressed in the Colloquium often differed from those then current in public debate on the Tunnel. Generally speaking it was assumed that the project was viable and would proceed so that the real problem was to control and exploit its likely effects. This was likely to be easier given a clear understanding of the project and of the systems in which it had to fit in the two countries and beyond. But, in line with the second aim of the volume, many questions remain unanswered about how control and benefit can best be achieved. International law, local planning arrangements, transport policy frameworks and questions of regional balance would all seem to have a part to play. The way all this works out will undoubtedly have a significant influence on the evolution of attitudes and therefore on European integration. Similarly, much remains to be done if both countries are to respond positively to the challenges of the Tunnel.

Monitoring of the project will thus require to be on a wide scale, paying attention to possible conflicts of laws, protection of investors, and the procedures for planning, co-ordinating site works and regional development. But such things need to be done positively and in a collaborative context. For what the Colloquium, and it is to be hoped, these papers, bring out is the way in which both sides aim at the same targets of protecting citizen and environment and maximising economic benefit, even though the problems they face and the way they go about solving them differ greatly and are not always well understood on the other side of the Channel. We hope that these papers and discussions will be the harbingers of more positive and understanding approaches to the future of the Channel Tunnel.

LEGAL AND FINANCIAL IMPLICATIONS OF THE CONCESSIONARY AGREEMENT

By Alain Fayard

After two unsuccessful attempts to create a permanent link between Great Britain and the Continent (in the 1870s and then in the 1970s), the French Head of State and the British Prime Minister announced that the project was being reconsidered at the Franco-British summit meeting of 10-11 September, 1981.

On 12 February, 1986 the two States signed the Treaty of Canterbury which provided for the construction and operation by private concessionary companies of a permanent cross-Channel link and, on 14 March, 1986 a quadripartite concession was signed between the two States, on the one hand, and France Manche and the Channel Tunnel Group, on the other.

The project calls on legal and financial techniques of a well-established nature, but has also required major innovations in both domains.

1. The Legal Aspects

In this case the two main legal systems which divide the Western world, have met face to face. The solution to the possible conflict has been found in creating a special body of regulations, within the framework of a concession, governing the relations between the States, on the one hand, and the concessionary companies, on the other, and in maintaining the application of national legal systems for the remainder. Moreover, a certain number of special regulations govern both the relations between the two concessionary companies and the relations between the latter and the lenders.

15

1.1 A French-English product : The Concession

The basic options of the concession were laid down by the guidelines contained in the invitation to promoters and by article 13
of the Treaty of Canterbury:

- a single concession for the whole of the project, deriving from
  international law, and which does not pre-judge the precise
  legal nature of the project;
- the absence of financial guarantees from the States;
- commercial freedom within the framework of normal regulations
  governing competition;
- income and expenses shared half each by the French and English
  concessionary companies;
- joint and several liability of the concessionary companies and
  a single executive body.

The text is quite clearly inspired by the concept of concessions
for public service utilities under French law (which is of
praetorian structure rather than a matter of statute law). Thus
it imposes a certain continuity of services and a power of
control by the Governments. However, it includes a number of
particularities, for example:

- a two-fold legal system with respect to land law matters which
  results from the profound differences between the national
  legal systems;
- the exclusion of the positive joint liability which, under the
  French system, binds the body granting the concession and the
  concessionary company and in particular the theory of lack of
  foresight or 'imprevision';
- the affirmation of the non-obligation for the bodies granting
  the concession to pursue the construction or the operation in
  the event of a default by the concessionary company.

An inter-governmental commission has been set up in order to
follow all questions relating to the construction and the
operation of the permanent link, in the name of the two

Governments and on their authority. A Safety Authority advises and aids the Inter-Governmental Commission and supervises the safety of the permanent link under the authority of the Inter-Governmental Commission.

Any possible disputes with respect to the application of the concession will be settled by a Court of Arbitration which will apply the relevant provisions of the Treaty and the Concession, any relevant principles of international law and any appropriate regulations under French or English law. And, where the parties agree, it will also apply the principles of equity.

1.2 Application of National Legal Systems

The concession which governs the relations between the bodies granting the concession and the concessionary companies expressly provides that the concessionary companies agree to respect the laws and regulations applicable at all times in each of the States including European Community law, and the provisions of the Treaty, complemented, where applicable, by additional protocols.

Disputes relating to the application of national legislation fall within the jurisdiction of national courts or of any other courts, where applicable, which are authorised by the national legal systems.

Concertation with local authorities and co-ordination of accompanying actions have been organised within the framework of national practices and laws. In France, the Delegation for land planning and for regional development (la Délégation à l'aménagement du territoire et à l'action régionale) assures flexible co-ordination between ministries. A protocol agreement between the State and the region Nord/Pas-de-Calais has been signed, together with additional clauses to the agreements of the State/Region plans involving the three regions involved. Finally, under the 'Grand Chantier' procedure an overall site co-ordinator is to supervise all operations related to the site, including housing, public equipment, social measures, vocational training and ensuring the participation of local and regional companies.

It is within this same framwork that the procedures for author-
isation of compulsory purchases and implementation of the Treaty
and the Concession will take place:

- In France, draft bills authorising the ratification of the
  Treaty and, where necessary, approving the concession, have
  been put before Parliament; their passage will automatically
  result in the applicability in French domestic law of the
  provisions of the Treaty. Moreover, a public enquiry and a
  joint investigation by the relevant administrative services
  have been carried out and a decree declaring the works to have
  a public purpose is before Conseil d'Etat, the highest administrative jurisdiction in France. Once this decree has been
  signed the State, and then the concessionary company, may
  proceed with compulsory purchases.

- In Great Britain a Hybrid bill was introduced before parliament.
  Those persons specifically affected have the right to register
  a petition in order to request Parliament to amend the draft
  bill; this petition is examined by the Select Committees of the
  House of Commons and the House of Lords. The passage of the
  bill will authorise compulsory purchases on the one hand, and,
  on the other the implementation of the provisions of the laws
  and regulations necessary for the application of the Treaty.

It therefore transpires that, over and above the differences of
procedures, the same aims of protecting the citizens, of
examining the public purpose of the project, and of introducing
the provisions of the Treaty as domestic law are being carried
out.

Finally, the juxtaposition of national controls which rely on the
establishment of the customs officials and police of one country
in the other country, raises problems of adaptation for British
law, although this practice is common on the Continent.

1.3   Adaptation of the national legal systems for the organisation
      of the concessionary companies and the protection of lenders

1.3.1   The concession obliges the concessionary companies to form
a common joint body entrusted with co-ordinating the whole of the

operations and of representing them vis-à-vis the Inter-
Governmental Commission. In the absence of legal instruments
which would establish them as a European company, the two
concessionary companies France Manche and Channel Tunnel Group
have erected a unified and integrated structure.

The two concessionary companies are respectively 100% sub-
sidiaries of the holding Companies Eurotunnel S.A. and EuroTunnel
plc and have created a joint venture company in EUROTUNNEL. The
shares of Eurotunnel S.A. and Eurotunnel plc are grouped in in-
dissociable units. The Eurotunnel company thus brings Frenchmen
and Englishmen together in one team led by a single general
management. Likewise the Boards of Directors have the same
members, subject to the national regulations on their composition.

1.3.2 Moreover, there is no a priori reason for introducing
provisions in the Concession in favour of third parties who are
lenders. However, the outcome of the concession agreement is
obviously not a matter of indifference to bankers since the re-
payment of their claims depends on its performance. The bank
syndicate would have liked to sign an agreement on this score
with the Governments. Given the refusal of the States to grant
a financial guarantee to the project, such a solution, whose
meaning would in any case have been ambiguous, could not be
entertained. The concession has, therefore, been chosen as the
legal foundation of an ad hoc system which is compatible with the
legal systems of the two countries. It aims to protect the
lenders while avoiding any dilution of liabilities vis-à-vis the
bodies granting the concessions, and while also assuring
reasonable continuity of the project. There are three components
to the system:

a)   Substitution makes it possible for the bank syndicate to
       take over the role of the concessionary company through the
       mediation of companies called "substituted bodies". This
       mechanism can be initiated either by the Governments,
       should they envisage the concession, or by the lenders
       provided they gain the prior agreement of the States.
       Substitution is not irreversible and must, in principle,
       terminate after the repayment of bank debts.

19

b)   In the event that the concession is cancelled, the prefer-
     ential right allows the banks to benefit from any new
     concession which might be signed.

c)   Indefensible rights (droit de suite) aim to allow the lenders
     to receive the appropriate payments from the net income from
     the Link, in the event that the operation is continued by a
     body other than the concessionary company vis-à-vis to which
     they are financially committed.

2.   Financial Aspects

One of the main principles laid down by the two Governments in the
guidelines issued to promoters in April 1985 was the exclusion of
any budgetary contribution and of any financial guarantee on their
part. Two constraints have resulted from this:

- the raising of an important sum of capital (roughly 20% of the
  finance requirements of the project) which assumes that the
  shares will offer sufficient return (of over 15% in the present
  state of the market);

- the implementation of a "financing project", which presupposes
  that the estimated revenue is sufficient to pay the interest on
  the debt with a reasonable safety margin. A cover ratio of 1
  to 2 has been accepted for the bank debt (ie, the relationship
  between the present value of the income and the outstanding sum
  of this debt); the ratio for the whole of the debt for the
  thirty-three year period of exclusive operation is therefore 1
  to 9.

The success of this financing relies on the adaptability of the
techniques of project financing and a correct evaluation both of
the risks involved and of their distribution.

2.1   The Adaptation of the Technique of Project Financing to the
         Financing of a Transport Infrastructure

A favourable context created by the widening of financial markets
and by innovations in financial instruments has facilitated this
adaptation. However, considerable special difficulties remain:

- The large size of the financing requirement which is of the
  order of 50 billion francs:

Works	27 billion
Overhead expenses	4 billion
Inflation	6 billion
Intermediate interest	10 billion

  Eight to ten billion francs will be covered by capital and 40
  billions by bank loans together with stand-by credits amounting
  to a further ten billion francs.

- Roughly seven years are required for the construction period,
  which increases the financing requirements and makes a loan of
  an exceptional length necessary. Since this will have to be for
  eighteen years refinancing, amounting to 25 billion francs, is
  provided for in the form of bonds issued as of 1995, ie, two
  years after the opening. In addition, no dividend on the
  registered capital can be collected before the eighth year at
  the earliest.

- The existence of political risks of delay, and a past history of
  successive failures of the project, which may deny it the
  confidence of financial circles.

It is, of course, necessary to take account of the positive
factors such as the very long life expectancy of transport infrastructure, and an income which, after all, is less unstable in
value than income from mining or energy. This is because it
derives from the traffic between France and Great Britain. The
two countries also represent an excellent political risk. While
the risk of tying up capital exists, that of insolvency can be
considered to be negligible.

Under these conditions, it has been possible to set up a financing
plan assuring a satisfactory distribution of risks between the
investors (capital) and the lenders (bank debt):

- the bank loan has made wide international distribution necessary
  in order to bring together the 40 banks figuring in the under-
  writing syndicate for an initial commitment ranging from 500
  million francs to 2 billion francs each. This distribution will

also be increased again with the syndication of a further 100 to 150 banks which should take place in the months to come and which should make it possible to limit the final contribution of each to roughly half of its initial guarantee. It should be underlined that the bank commitments shall only become final on the signature of the credit agreement, saving only any possible suspensive conditions.

Since the loan is mainly made out both in French francs and in pounds sterling, various techniques must allow the banks to intervene in other currencies and to limit the exchange risks. Finally, the right of substitution, provided by the Concession (cf. 1.3 above), added to the usual pledge of shares, makes it possible to give the lenders the equivalent of a real right over the installations which is customary under project financing, but which was incompatible with the public property nature of the undertaking.

- the first issue of capital of 460 million francs was raised amongst the founder shareholders in order to cover expenses till November 1986. The second issue of 2,060 million francs was placed privately in October 1986 with institutional investors (35% for France, 35% for Great Britain, 10% for Japan, 8% for North America, 4% for Germany and 8% for the rest of the world). This will cover expenses until October 1987. The third issue of 7·5 billion francs, a figure liable to be reduced to 5·5 billion, will be realised in the form of a public issue after the ratification of the Treaty and the implementation of the Concession. Two thirds will be paid on subscription with a further sixth being payable in both 1988 and 1989.

2.2 Evaluation and Distribution of Risks

The success of this financing does, of course, assume that the risks have been correctly evaluated and that their distribution between the various participants is satisfactory.

2.2.1 Evaluation of Risks

The evaluation of the risks is an absolutely decisive factor since the greater the uncertitude, the higher the cost of financing through interest rates, the size of shareholders' equity and cover ratios.

a)   the construction risks (non-completion, overstepping of costs and extension of delivery dates) are limited by the use of the technique of the drilled tunnel and the recommencement of the already very elaborate project abandoned in 1974. However, while the risk of non-completion can be considered to be insignificant, on the contrary the risk of overstepping the costs and, above all, of extending the delivery dates is far from negligible, even if the stand-by credit makes it possible to cover up to 10 billion francs of this. This is all the more so given that, in order to reduce the time necessary for completing the project, and thus the finance required, techniques which have never been used under the extreme conditions envisaged even if they are essentially well-known. For example: boring machines will be working under pressures of seven to ten bars and the absence of intermediary borings could block the work site in the event of delays in breaking through a geologically difficult zone.

b)   The operating risks appear to be significant:

- Traffic estimates have always been a particularly delicate art; and in this case they can only be guessed at since details of the service to be offered are, in reality, still unknown. The frequency, safety, comfort, speed and stability of the shuttles, and the length of time to be spent in loading and shuttling are still open questions. There is also the question of the psychological response to the fact that there is still no drive-through link to be remembered.

- So although the group of five credit arranging banks which presented a report to the Government in May 1984 drew up estimates of traffic, the highest estimate of which corresponded to the medium estimate envisaged by the Franco-British study group in 1982 these are still very provisional. Thus in October 1986, these estimates were 40-80% higher for passengers (with or without the high-speed train (TGV)) and 60% higher for freight. By March

1987, Eurotunnel was announcing a mark-up of 10-15% on
these figures and June 1987 saw further upward revision.

- Income is also extremely sensitive to the commercial policy
  followed by the ferries. Exemption from V.A.T. has not yet
  been accepted, and will be decided by the European
  Community and not by French and English Governments.
  Commercial freedom and the non-applicability of any
  possible price controls, granted in consideration for
  private financing, should not be allowed to mask the fact
  that national and EC regulations relating to competition
  and to abuse of dominant position will be applicable.

- Similarly the costs of operation are, of course, still
  badly defined and the safety regulations on loading, fire
  doors, and decisions as to whether passengers can travel in
  immediate proximity to their vehicles or not, could result
  in very high costs, in addition to their influence on the
  commercial attractiveness of the link.

c) Financial risks will be largely covered once the bank credit
has been set up and the capital has been issued. However,
exchange risks, despite hedging agreements and swaps, may be
significant and a cycle of less favourable inflation than the
estimate of 3·5% in 1986 and 6·0% in 1991 would have negative
consequences.

2.2.2 Distribution of Risks

But more than the evaluation of risks by the market, even if the
large intervention of semi-nationalised investors or the pressure
of a central bank can be of a nature to somewhat bias the judge-
ment of market makers, it is the distribution of these risks which
is primordial for fully understanding the reality of the wholly
private nature of the financing and its total lack of public
guarantees.

a) The distribution of risk between investors and lenders
results from the distribution between the shareholders'
equity on the one hand and the bank credits and securities
from which the lenders benefit on the other.

b) The construction agreement has highlighted the existence of potential conflicts of interest for the contractors (and, indeed, for the banks) who have promoted the project, between their roles as initial shareholders and their roles as future suppliers of services.

It is possible to imagine that the works agreement, and the loan agreement, would only be confirmed after the implementation of the concession. In fact, it was signed with Transmanche Link, a consortium grouping the ten original shareholding contractors, on 13 August, 1986. The release of capital took place in October 1986. In accordance with the Declaration by the Representatives of the Governments of the Member States, meeting in the Council of 26 July 1971, this contract was settled without competition or publication, which is the counter-balance of the fact that the risk is borne by the contractors as partners of the concession. The agreement is in three parts:

- lump sum works for the terminals and equipment of the tunnel, roughly 10 billion francs

- cost plus fee works with a target price for the tunnels, roughly 12 billion francs

- supplies contracts for the rolling stock, roughly 3 billion francs.

The contractor, therefore, takes the risk for the terminals and the equipment. However, with respect to the drilling of the tunnel, he is repaid for his real expenses plus a fee of 12·36%. Where the target-cost is overstepped, he pays 30% thereof, up to 6% of the target-cost. In the event that the real expense is less that the target-cost, he enjoys a 50% bonus on this saving. For supplies, the contractor collects a fee of 11·5%, but is not responsible for the cost risks.

The appropriateness of these estimates has been examined by the independent 'maître d'oeuvre' provided by the Concession and who can play the role of technical expert on behalf of the bodies granted the concession. This 'maître d'oeuvre' stated that the "Pricing is reasonably in view of the

position of the contractors in promoting the project". This
does not appear to reveal perfect clarity about the
distribution of risks and roles.

However, it is fair to note both that the works agreement was
the subject of a minute examination by independent experts
appointed by the new shareholders and that a management independent of the founders and shareholders was set up.

c)   Finally, the agreement for the use of the fixed link by the
railway companies which should contribute more than one third
of income, may be of importance in the distribution of risks.
This is not only due to the relations between public
authorities and the companies but, above all, due to problems
of co-existence between shuttle traffic and direct trains.
Too high a toll for railways companies might be thought as a
cross subsidy which would not be lawful in view of
competition regulations.

In addition, the problem arises of the eventual construction
of a Paris/Brussels/London very high-speed train (TEV), a
project to which the concessionary companies seem to attach
much importance, less apparently for financial reasons than
as a demonstration of the commitment of the public
authorities. A provision of the agreement with the railway
companies, whose commercial logic is difficult to grasp,
provides that the tolls shall be higher for ordinary trains
than for high-speed trains and the loss of income is therefore relatively slight.

Financial immunity for the States therefore risks being
poorly assured. These appear, via certain interventions of
railway companies, as the ultimate recourse which can allow
the financing of a project where the risks are thrown onto a
concessionary company, which is not the real promoter, and
which has a significant but limited financial status. In the
event of difficulties, the project's nature as a public
utility, on the one hand, and the risk of undermining the
credit of the two countries, on the other hand, given the
precedents of the motorways in France and the present unease

due to the default of a Catalan electrical company, could force the States to intervene.

In all events, the setting up of the permanent cross-Channel link shows the wide possibilities of adaptation of the legal and financing implements both on the Continent and in Britain and may serve as a basis for projects, either with totally private financing or with mixed financing.

References

Emmanuel Guillaume : 'Un arbitrage réussi' in Revue P.C.M. 1986, No. 10

Pierre Mayer : 'Réflexions sur le financement privé de liaison fixe transmanche' in Schweizer Journal 1/1987

Bernard Thiolon : 'Le financement privé du tunnel' in Revue E N A No. 168, January 1987

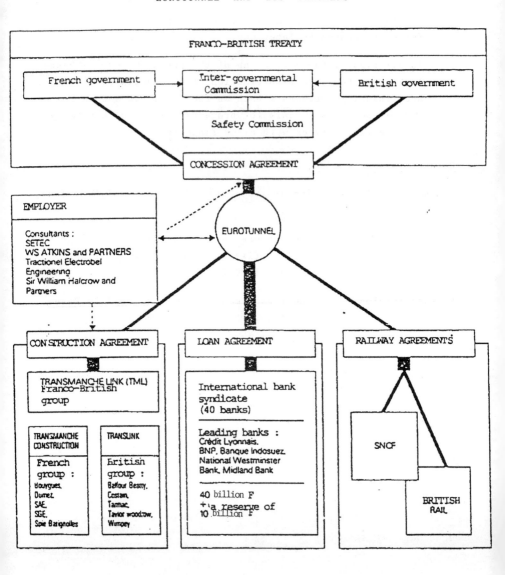

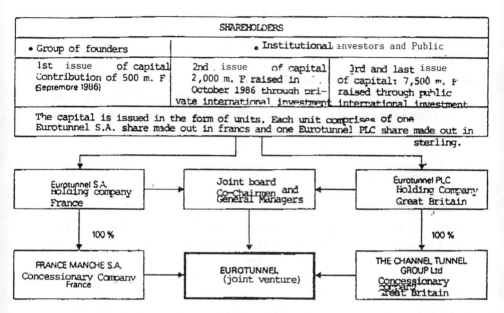

SHAREHOLDERS		
• Group of founders	• Institutional investors and Public	
1st issue of capital Contribution of 500 m. F (Septembre 1986)	2nd issue of capital 2,000 m. F raised in October 1986 through private international investment	3rd and last issue of capital: 7,500 m. F raised through public international investment

The capital is issued in the form of units. Each unit comprises of one Eurotunnel S.A. share made out in francs and one Eurotunnel PLC share made out in sterling.

Eurptunnel S.A. Holding company France → **Joint board Co-Chairmen and General Managers** ← **Eurotunnel PLC Holding Company Great Britain**

100 %   100 %

FRANCE MANCHE S.A. Concessionary Company France → **EUROTUNNEL (joint venture)** ← **THE CHANNEL TUNNEL GROUP Ltd Concessionary Company Great Britain**

The participants in the Workshop considered that the preparatory
"position paper" set out the general legal and financial problems
in a reasonable manner. However, it set itself to go more deeply
into a certain number of key points.

1. International Law

Dr John Kish enlightened the participants on the problems of
international law, notably delimitation, jurisdiction,
ratification, consular protection, arbitration and defence. The
exact delimitation of the boundary between the French and British
shore of the continental shelf remains to be decided, while there
are also arrangements for the exercise of sovereignty and the
circulation of officials to be negotiated. Many of the provisions
of the treaty have, however, been anticipated thanks to an
exchange of notes, so that they did not have to wait for formal
ratification. Other legal instruments like the 1951 Franco-
British convention and the North Atlantic Treaty may also be
applicable in certain circumstances. It thus emerged that the
Treaty of Canterbury is a framework which ought to be elaborated
by specific protocols and agreements as is, of course, already
envisaged in respect of general rules of international law.

2. The Concession and Intergovernmental Commission

Professor Gerard Marcoud emphasised the specific characteristics
of the system introduced for the Channel Fixed Link. In
particular the problem of knowing whether the concession creates
obligations between the states or whether all obligations between
the states are limited to those in the Treaty has to be resolved.
On the one hand the Intergovernmental Commission seems to be an
international authority, without a moral personality; its
composition and way of working make it an organ of a diplomatic
character. Nevertheless, it also has certain powers which will
produce changes in the internal legal system of each State. The

question is thus to know how its decisions will be incorporated into this internal system.

3. Application of National Law

The working party considered that, where the application of . national laws to the problem of planning and the environment was involved, the matter would be dealt with by the Workshop on "Planning and Environmental Aspects".

With regard to the problem of guarantees (article 31 of the Concession) and to substitution (article 32) the Workshop stated that the mechanism of guarantees appeared to be subsidiary to the arrangement of substitution. This was all the more so because of the principals of public domainality in France and the rule in the concession, according to which the execution of guarantees must not affect the carrying out and running of the project, all of which greatly limited the effects of these guarantees. With regard to the arrangement of substitution it was not possible to examine it in depth. However, questions arose, notably on relations with banks, because of the impossibility in English Law of making rules to cover foreign bodies.

4. Competition

The Concessionaires can fix their tariffs freely but are subject to appropriate national and Community laws about competition and abuse of a dominant position. Here there seems to be a potential for conflict between the different legal provisions, as was under-lined by Professor John Craven. Thus British authorities such as the Office of Fair Trading, the Monopolies and Mergers Commission, or the Secretary of State may impose modifications of tariffs in cases of anticompetitive practices. In France the powers of the Conseil de Concurrence and other jurisdiction may also impose sanctions.

The participants also questioned the risk of cross subsidisation mainly on the use of the tolls to be paid by the rail companies and the risk of intervention by public powers in the name of national security or in the case where the Tunnel found itself in grave financial difficulties.

5. Financing

Charles MacKay stressed that the economic and political context
had become particularly favourable and that the method of
financing the project (Project Based Financing) representd a
significant evolution because it allowed banks to grant loans
without having proper legal remedies. This was because the
project constituted a viable economic unity susceptible of
generating sufficient income to cover both running costs and the
servicing of the debt, and thus with a reasonable safety margin.
This method did not represent a new source of finance but a method
of financial technique which allows a division of the risks
between the lender and the borrower. However, the use of this
method is evidently not compatible with the search for the lowest
possible price, because taking risks has its price.

The participants also stressed the profound difference between
this project and the logic of financing found in the 1974 project,
where loans were largely guaranteed by the State and where there
was an actual guarantee of income. If the risks of non-completion
really appear negligible, on the other hand the assessment of
other risks is rather tricky. The use of techniques which may
permit rapid boring but which have never been used in the extreme
conditions now envisaged, the limited knowledge of the service to
be provided by the Tunnel, involving an operating mix of shuttles,
and high speed trains (whether with an efficient connection to
London or not), and the quality of treatment of goods trains,
passenger trains or car-sleepers transferring from continental
gauge at the British Terminals at Cheriton or Ashford, all remain
open questions.

Also the project requires a large volume of capital because it is
a project 'sans passé' differing from most other methods of
financing projects. Thus the costs of oil exploration are
supported by existing oil companies, the bridge project at
Dartford will be covered by the existing tunnel company, but there
was no existing body on which the tunnel could draw.

****

This brief account does not pretend to recreate the whole of the work of the Workshop, nor all the interventions of various participants. However, it should be noted that Professor Christophe Dupont presented the research undertaken on aspects of the negotiation of the Transmanche project, while that M. Daniel Ghouzi underlined the regional aspects of the Channel Tunnel. Particular thanks must also go to M. Michel Gentot who allowed other participants to make particularly instructive comparisons with the 1974 project on which he actively collaborated.

PLANNING AND ENVIRONMENTAL CONCERN IN BRITAIN

By Margaret Anderson

There are two major reasons why the planning and environmental
aspects of the proposed Channel Tunnel have aroused debate in
Britain. These are:

(1)   that normal planning procedures have been 'ignored' in
      favour of a parliamentary procedure which many feel will
      not allow a full enough exploration of all aspects of
      the project;

(2)   that the environment of South East England, and of Kent
      in particular, will suffer not just from the tunnel and
      its associated works but also from inevitable secondary
      developments.

This paper looks at the background to these two issues.

1.   PLANNING   CONSIDERATIONS

1.a Normal planning procedure - a public inquiry

Since 1947 all development in Britain has required permission
which is normally obtained by the developer making an application
to the local planning authority. Where a major application has
more than local implications or is likely to arouse more than
local opposition the Secretary of State (SoS) for the Environment
can 'call in' the application. This procedure enables the SoS to
remove the application from local jurisdiction, but he is then
required to hold a hearing, or public inquiry, before a decision
on the application is reached.

Public inquiries are held before an Inspector appointed by the
SoS. Unlike a court of law, evidence is not usually given on oath
and the Inspector has discretion to admit any evidence, even

hearsay evidence, which he considers to be pertinent to the application. The main bulk of the evidence will cover three main areas:

 (i) the facts - which may be disputed by either side

 (ii) expert or technical opinion brought in to reinforce matters of fact

 (iii) future intentions - which may be ancillary to the main intention of the applicant but which may affect the future use of the site, or may refer to the future activities of the local authority.

At the inquiry the applicant (or his advocate) usually presents his case first, followed by the local authority and then by other parties who have registered their interest. Witnesses may be called and may be cross-examined. After the close of the inquiry the Inspector writes a report summarizing the arguments and making a decision recommendation to the SoS. The final decision rests with the SoS.

Public inquiries into major developments have recently been heavily criticised particularly on two counts: that individuals and small groups are at a considerable disadvantage faced with the legal and financial power of the large developer; that the inquiries are inordinately lengthy. Both these problems were exemplified by the inquiry into the nuclear power station at Sizewell which took more than four years from the start of the inquiry in January 1983 to the final decision in March 1987. A Channel Tunnel inquiry would undoubtedly take at least as long.

Despite these acknowledged problems, opponents of the Channel Tunnel have consistently called for a public inquiry in the face of the government's decision to proceed instead with a hybrid bill in Parliament.

1.b Channel Tunnel procedure - a hybrid bill

One reason for the government's decision to use the hybrid bill procedure was the perceived need to keep as closely in time as possible with concurrent procedures in France. However, the government's sense of urgency runs counter to what the public sees as the need for a fair hearing.

A hybrid bill is so called because it contains 'provisions not only of public law, but of private law directly affecting individuals' (Cmnd 9735 para 65). A hybrid bill not only goes through all the normal Parliamentary stages but is also examined by a Select Committee in both the House of Commons and the House of Lords. All those 'bodies and persons whose interests are directly affected may petition and have their cases heard' before the Select Committees. 'The right to petition depends upon the standing (*locus standi*) of the petitioner - a matter which is determined by each Committee in accordance with the practice of Parliament' (Cmnd 9735 para 65). In this case the Government, as sponsor of the bill, undertook 'not to seek to oppose the right of anyone to appear before the Committees on a petition to secure protection, either for their personal interests, or for the proper interests of any organisation or group which they may have been appointed to represent' (Cmnd 9735 para 65).

It may be thought that both the procedure of Committees and the government's undertaking would have satisfactorily dealt with objections to the refusal to hold a public inquiry. However, there are a number of aspects of the hybrid bill system which the petitioners see as especially disadvantageous in this case. In particular, once the Bill has been formally accepted into the House of Commons discussion must be confined to the actual content of the Bill. Petitioners may not therefore raise other points, such as the need for a fixed link, nor may they continue to press for a public inquiry. In addition, the petitioners felt the time-scale was far too short and feared that the speed of the procedure would allow domination of Committee time by the major participants and that other legitimate interests, particularly those of local individuals and of the environment, would be squeezed out.

One of the arguments put forward by petitioners in favour of a public inquiry was that they would then have been able to argue the principle of a fixed link and not just be confined to detail. In fact, however, there is no guarantee that the need for a development will be considered a legitimate item for discussion even at a public inquiry. The main advantage of an inquiry for those opposed to the Channel Tunnel would probably have been the possibility of using it as a delaying procedure with the hope that, as a result, the project would fall. On the other hand a major advantage of the hybrid bill procedure is that it is possible for 'bargaining' to take place and for petitioners directly to obtain concessions in design or construction of the development and to obtain guarantees that certain things will (or will not) be done. In many ways the promoters and operators have to be more accountable in Select Committee hearings than at a Public Enquiry. In the former, members of the Committee can insist on detailed answers to questions, the provision of additional information and alterations to the scheme; in the latter it is easier to evade questions and the role of the Inspector is to listen, record and ultimately make a recommendation.

The Channel Tunnel Bill was given its Second Reading in the Commons on 5 June 1986. The Commons Select Committee was appointed 16 June 1986, and 4845 petitions had been received by the closing date, 27 June, the majority from individuals. The Committee hearings began on 24 June and continued on 36 days up to 6 November. The deliberate decision by the government not to challenge the *locus standi* of any of the petitioners, while laudable in that no one need have felt excluded, may have been counter-productive because many petitioners did not fully understand the limitations of the hybrid bill procedure and were therefore frustrated and aggrieved if the points they wished to raise were ruled out of order.

1.c   Impact of the Parliamentary procedure on Local Authorities' planning responsibilities

Two main aspects of the planning responsibilities of Local Authorities are affected by the Channel Tunnel Bill procedure:

(i)   County structure plans and District local plans

(ii)   development control.

1.c.i   Structure and Local Plans

Each County Council (eg, Kent, East Sussex) is required to prepare

37

a comprehensive, long-term plan for its county - a structure plan.
Structure plans set out, and justify, general proposals for
development against a background of national economic, social and
environmental policies. They consist of a written statement and
diagrammatic material but no detailed maps. As such they provide
the framework for the local plans which may be prepared by each
District Council (eg Ashford, Dover) within the county. Local
plans may contain policies for a particular area or town (eg
Folkestone) or a subject (eg recreation); they are based on
detailed maps.

The public are involved at various stages in the preparation of
structure and local plans. Structure plans have to be approved by
the SoS; local plans must conform to the relevant structure plan.

The Kent structure plan, prepared in the 1970s made provision for
land to be safeguarded in the area of Cheriton for a Channel fixed
link terminal. This area is delineated in the Shepway District
local plan. The structure plan also designated Ashford as a
growth point for Kent. When the fixed link project was revived
Kent structure plan and several District local plans were already
under review as part of the normal, rolling programme of plan
review. These reviews will now have to anticipate the
developments likely to stem from a successful tunnel project and
plan for their location and development.

1.c.ii   Development Control

All development in Britain requires permission, which may be
obtained in a number of ways the most usual being, as already
mentioned, through an application to the local authority. However,
in certain instances permission may be 'deemed to have been
granted' when the development forms part of an agreement reached
under some other government procedure. 'Deemed permission' does
not mean 'carte blanche'. The agreement will contain closely
specified detail on points which must be submitted to the local
planning authority before development can begin.

In the case of the Channel Tunnel, any of the developments
detailed in the hybrid bill and subsequently contained in the Act

will have 'deemed permission', but any developments that are not included in the final Act will have to go through the normal planning procedures. Thus it is vital for the operators (Eurotunnel) that they think of every eventuality and make sure it is included; that the government (or any petitioner) tries to include any development, such as roads, that it considers essential for the immediate operation of the tunnel; and that petitioners also try to get excluded from the Bill any developments which they wish to ensure will go through the normal planning procedures including, perhaps, a public local inquiry. Thus, for example, the government included in the Bill provision for improvements to the A20 trunk road leading eastwards from the terminal site towards Dover, and Kent County Council petitioned for a visitors' viewing area at Cheriton and consequent improvements to access roads to be provided by the operators.

2.   ENVIRONMENTAL   CONSIDERATIONS

2.a   Environmental protection zones in South East England

The South East of England, and Kent in particular, is acknowledged as an area of high landscape quality. The need for protection of its farmland, settlements and woodlands and its natural geological and biological features has been specifically recognised in the variety and number of statutory and non-statutory designations that occur in the area. The attached map (Fig 1.) demonstrates the extent to which these measures cover the south east region and constrain development.

2.a.i   Statutory designations

    - Areas of Outstanding Natural Beauty (AONBs)

Designation as an AONB gives statutory recognition to areas of national landscape importance. Boundaries of AONBs are defined by the Countryside Commission and the designations confirmed by the Secretary of State for the Environment. There is normally a presumption against development in AONBs with particular policies being spelled out in structure and local plans.

The south eastern end of the Kent Downs AONB will be the most affected by the proposed tunnel (see Fig 2.) and, although the terminal site at Cheriton, and the adjacent Castle Hill and Sugarloaf Hill are not included in the AONB they nonetheless form an integral part of its landscape setting and their development gives rise to outspoken objections. Kent Downs AONB ends abruptly in the chalk face of Shakespeare Cliff - the well-known 'White Cliffs of Dover' which holds a strong place in the hearts of the British people. The proposals that may disrupt or destroy them raise considerable alarm.

- Green Belts

Green belts are designed to check urban sprawl; their boundaries are defined in structure and local plans. Despite attempts in recent years to weaken green belts the government has reaffirmed its intention (1984 and 1987) not to allow development in green belts except in very special circumstances.

London's green belt was defined in the 1950s and has gradually extended until it is now 20-25 km wide in most places. The building of the M25 circling London through the green belt has increased pressure for development particularly at interchanges. The projected rise in cross-Channel traffic in the coming decades, whether through the tunnel or the ports, will intensify this pressure.

- Sites of Special Scientific Interest (SSSIs)

SSSIs are established by the Nature Conservancy Council (NCC). They are areas of special importance because of their flora, fauna, geological or physiographical features. Designation of these sites is by statutory notification by the NCC to the local planning authority, landowners and occupiers and the SoS. If local authorities plan to allow development to take place on the site they have a duty to inform the NCC who may make represent- ations under normal planning procedures.

There are two SSSIs within or close to the proposed tunnel working. Folkestone Warren SSSI is a Grade I site of international

importance. It is an area of undercliff resulting from landslips
and contains a diversity of coastal habitats. The area is also of
geological interest. The construction activity and the deposition
of spoil behind a retaining wall at Shakespeare Cliff are not
likely to cause direct loss of habitat in the short term, but in
the long term the effects of changes in current and wave patterns,
resulting from the extended working platform, will almost
certainly have an effect on the distribution and variety of plants
and animal life.

The Folkestone to Etchinghill Escarpment SSSI is a Grade II site
of chalk grassland and scrub within which is Asholt Wood ancient
woodland (Grade I). There is a proposal to extend this SSSI to
include Holywell Coombe which is of particular geological interest.
There is also a proposal for Seabrook stream, which runs to the
west of Cheriton, to be designated an SSSI; parts of its upper
reaches are already protected as they lie within Asholt Wood. As
currently proposed, the road and other works associated with the
western end of the terminal will skirt Asholt Wood but cleaned
surface run-off may be diverted into a lagoon and thence into
Seabrook Stream. There is very considerable disquiet over the
likely polluting of the stream. The other main areas of concern
are Castle Hill where the tunnel begins and the cut and cover
tunnel at Holywell Coombe which will also be crossed by a viaduct
carrying the A20 road improvement from Folkestone towards Dover.
It is accepted that a certain amount of environmental damage is
inevitable in a major project but the NCC and others are anxious
to ensure that Eurotunnel maintains its pledge to strict
monitoring and careful management.

2.a.ii Non-statutory designations

    - Heritage Coast

The Countryside Commission introduced the idea of Heritage Coasts
in 1970. They are defined as coastline of national and scenic
importance and their designation is intended to draw attention to
their importance with a view to resolving conflicts over their use
for recreation, farming and intensive development. Boundaries of
Heritage Coasts are defined by local authorities and policies for

their protection set out in structure and local plans. (Figure 2)

It is not very surprising that the White Cliffs at Dover, including Shakespeare Cliff where the Channel Tunnel working platform is to be extended, is Heritage Coast. The designation continues westward to embrace also the Folkestone Warren SSSI.

- Kent Countryside Plan policies (Figure 2)

Kent County Council has prepared a local subject plan for the countryside. In this it defines areas of special significance for countryside conservation in order to permit interested parties (eg developers) to see more readily where there are likely to be particularly stringent constraints on development (Kent CC 1980). Several of the areas defined overlap each other and some do not completely coincide with the area statutorily defined. For example, the Special Landscape Areas cover land beyond the boundaries of the AONBs.

The countryside conservation areas are:

- Areas of Special Significance for Agriculture: ie the most extensive tracts of grades 1 and 2 land;

- Special Landscape Areas, of which there are eight: these include the three AONBs and the Heritage Coast and areas of countryside which provide an unspoilt foreground to the more striking scenery;

- Areas of High Conservation Value: these comprise extensive tracts of those habitat types or physiographical features represented in Kent which are most rare and sensitive to change;

- Undeveloped Coast: definition of these areas is designed to protect the unspoiled appearance and scientific importance of the undeveloped shoreline not already designated Heritage Coast.

All these areas are subject to Kent's structure plan policies. As far as the Channel Tunnel is concerned the terminals and construction sites will impinge on all types of conservation area except that for agricultural land. For this reason a considerable proportion of the effort of petitioners against the Bill has been concentrated on ensuring mitigation of the environmental damage, monitoring and restoration.

Equally, much of the anxiety about future development that may stem from a successful tunnel enterprise contres on the further destruction of Kent, the 'Garden of England'. It is clear from the map (Figure 2.) that only the band of the Low Weald between Ashford and Edenbridge is clear of these specific environmental policies. Nonetheless it is land that no one would want to see 'covered in concrete'. A fate some people anticipate may be the future for large parts of South East England.

2.a.iii The Built Environment

Part of the charm of Kent lies in the long history of its settlements, in mellow brick and tile houses, spectacular public buildings such as cathedrals and castles, and unique features such as Kentish oasts. Much of the built environment blends with the natural environment to give Kent its special character.

The main measures available for protection are the listing of buildings of particular historic or architectural interest, and the designation of urban conservation areas. In both cases protection is given against demolition or alteration and grants are available for repairs. In conservation areas designation enables greater attention to be paid to the total scene including appropriate street furniture, signs and lighting. An indication of the wealth of attractive and historic towns in Kent is given by the fact that there are 40 urban conservation areas in Dover District alone.

The proposed construction and permanent works for the Channel Tunnel will have little direct effect on the main adjacent centres of Folkestone and Dover. However, a number of small settlements will be severely affected, most notably the villages of Danton Pinch, Newington, Peene and Frogholt which lie within or on the western edge of the Cheriton terminal site and its associated approach roads. Fifteen properties will be destroyed. Disruption will clearly be very severe for all the inhabitants of

these villages, some 240 people, and to the few who live in the
immediate vicinity of the construction works at Shakespeare Cliff.
It was estimated by Channel Tunnel Group in 1985 that potentially
a further 5,000 people could have their lives affected to some
extent by the tunnel development.)

    - Ashford

Ashford is designated a growth area in the Kent structure plan.
It lies in an area of little environmental constraint and plans
for development along the south east side of the town are included
in the local plan.

The Channel Tunnel operations away from Cheriton centre in Ashford
(Figure 3.). The international railway station for passengers
will be built on existing British Rail land adjoining the present
station where five lines meet. Access to the station from the
south will require a new road and some demolition of existing
buildings. The Inland Freight Clearance Depot will be built on
the south east edge of the town, adjacent to the railway line and
reached via a direct road link from the M20. This road, the South
Orbital Road, is being built to take traffic from the M20 past the
Clearance Depot and new Business Park through to link up with the
new access road to the International Station, and parking for
5,000 cars. This will keep the expected major increase in cross-
Channel traffic away from the centre of the town, which is a
conservation area.

3.   CHANGES TO THE CHANNEL TUNNEL BILL

The House of Commons Select Committee reported in November having
made 70 amendments to the Bill in the light of evidence from the
petitioners. In some cases it was decided it was inappropriate to
amend the Bill and points by petitioners were dealt with by
obtaining assurances from Eurotunnel, or the government, which
were agreed by the petitioners. No amendment to an assurance,
arising from subsequent debate on the Bill, will be made without
renegotiation with the relevant petitioners. The Select Committee
stated, 'We believe that great reliance can be placed on
assurances given formally to the Select Committee'. (Select

44

Committee Report para 6).

The Bill as amended by the Select Committee was debated by
Standing Committee A of the House of Commons over ten days from
December 1986 to January 1987 and by the whole House on 3 and 4
February 1987. Further amendments were made at both these stages.
The Bill received its Second Reading (formal acceptance) in the
House of Lords on 16 February. It was then considered and amended
by a House of Lords' Select Committee between 2 March and 30 April
1987 after hearing further petitions. The changes were agreed,
along with some Government changes, by the Committee of the House
between 2 and 6 July, with the report stage following on 13 July.
The Bill was passed by the Lords on 16 July. The Commons voted to
accept the Lords' amendments on 21 July and the Bill completed its
passage through Parliament a few days later, with Ratification of
the Treaty taking place in Paris on 29 July 1987.

REFERENCES

Ashford Borough Council (1986) Ashford and the Channel Tunnel,
The Council, Ashford, Kent.

Channel Tunnel Group (1985) The Channel Tunnel Project
Environmental Effects in the U.K., The Channel Tunnel Group,
London.

Department of the Environment (1984) Green Belts, Circular 14/84.

House of Commons (1986) Special Report from the Select Committee
on The Channel Tunnel Bill, H.M.S.O., London.

Kent County Council (1980) Kent Countryside Plan, The County
Planning Department, Maidstone, Kent.

Figure 2 is reproduced with permission from the Kent Structure
Plan (Maidstone, Kent County Council, 1981).

Figure 3 is reproduced with permission from Ashford and the
Channel Tunnel: a brief introduction to Ashford's role in the
project (Ashford Borough Council, 1987).

Figure 1.   MAJOR CONTRAINTS ON DEVELOPMENT IN
SOUTH EAST ENGLAND

 Metropolitan Green Belt

Areas of Outstanding Natural Beauty

● National Nature Reserves

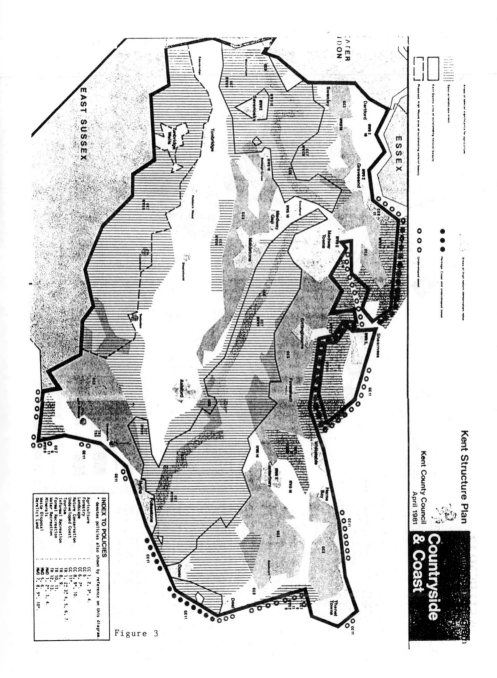

Figure 3

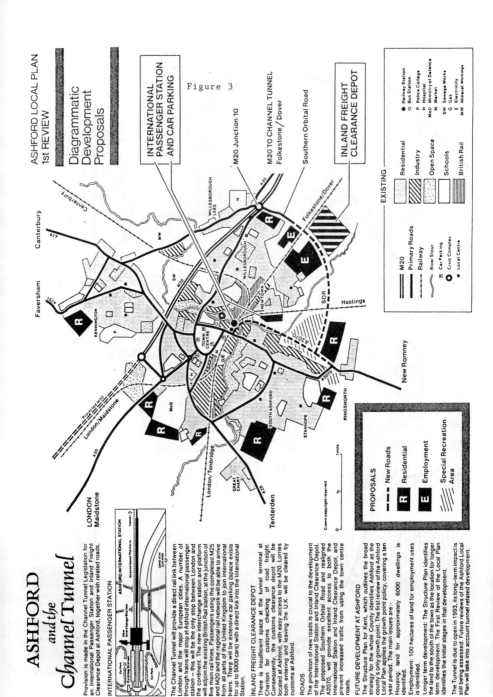

Figure 3

----

1.   The background paper for this workshop

Considered the two main reasons why planning and environmental
aspects of the Channel Tunnel have aroused debate in Britain.
These are:

(a)   The planning procedure used.   Instead of the tunnel
development going through the normal British planning system which
would mean with a project of this magnitude that the case would be
heard at a public enquiry, the government decided to promote a
Hybrid Bill in Parliament.   This is a quicker method, and does
still allow organisations and individuals to argue details of the
project and, very importantly, it does enable changes to be made
to the design, organisation and operation of the proposed project
during the hearings.   However, the principle of the project may
not be considered.   Despite all the acknowledged disadvantages of
a public enquiry many people felt strongly that there should have
been one.

(b)   The environmental consequences.   Kent is known as 'the garden
of England' and fears are very strong that not only will nature
conservation and landscape in the immediate vicinity of the tunnel
workings be ruined, but that consequential development arising
from the success of the tunnel in the future will be detrimental
to the County, despite strong protectionist policies.

2.   Against this background the workshop group decided to look
particularly at the parallel French aspects of these topics and
considered:

(a) the French procedure for promoting the tunnel;

(b) the environmental problems on the French side;

Following these the group then discussed:

(c) the wider strategy implications of the tunnel in Britain,
    France and other European countries;

(d) the organisation of major construction sites.

Taking these in the above order:

2(a) <u>Procedure</u>

It was agreed that both countries have had a problem of dealing with a private project inserted into the public system.

The normal French procedure is for the public to give evidence on a project before a local Commission d'Enquête, which is an independent committee created by the Prefet. The Commission then submits its report and advice to the Conseil d'Etat and the decision is then made by the appropriate Minister on the advice of the Conseil d'Etat. At the same time there is a parallel central procedure of consultation between ministries and government bodies. This normal procedure was followed for the Channel Tunnel.

Expropriation of property is only possible for a project which has been declared to be in the public interest, and is carried out by a duly authorised judge and accompanied by fair compensation. The declaration of public utility is only conceded, furthermore, when the investigation concludes that the balance of advantage clearly outweighs environmental and other disadvantages. In the case of the Channel Tunnel, the Prefect of the Pas de Calais, acting for the Ministry of Equipment, held the necessary enquiry in June and July 1986. After the Commission of Enquiry concluded in favour of the project the Prefect, over the winter, drafted his views on the value of the project and the consequent changes to local plans. And, at the same time, the Direction des Routes conducted an assessment from the point of view of the centre. This was then sent to the Conseil d'Etat for final approval which was given on 14 April 1987 allowing the declaration to be made on 8 May 1987. The terms of this are binding on the concessionaire.

The group found it interesting that, although the British government see the Tunnel part of the project as number one priority, and the rail and road infrastructure as secondary, the Channel Tunnel hybrid bill did contain important clauses on the <u>directly</u> related roads and railways. All other development, including infrastructure, not included in the Bill, will have to go through the normal planning processes, including, if necessary, public inquiries.

The French, on the other hand, see the completion of the TGV as the first priority of which the Tunnel is an inter-dependent part. But, there will have to be three separate Commissions d'Enquête, for the Tunnel, the TGV and the roads, of which only the first has so far been held.

It was felt that the parliamentary procedure in Britain had considerable advantages over the public enquiry because (i) evidence is given and petitions made by the public to select committees in both the House of Commons and the House of Lords and in both cases the Committee ask searching questions of the promoters and operators and insist on answers; it is easier to evade direct questions at a public inquiry; (ii) it is possible to change parts of the scheme by agreement between the promoters, operators and petitioners. Thus some real progress can be made in meeting objections during the process. Public inquiries and the Commissions d'Enquête, on the other hand, are only consultation procedures.

2(b) The Environment

It was agreed that for topographical reasons alone the British have considerable problems in locating and designing the various parts of the tunnel project. The outcome means using three sites some 30 km apart (from Shakespeare Cliff to Ashford); there is also the need to make as little disturbance as possible to an area of acknowledged landscape and conservation interest.

The French have the advantages of a larger site where all facilities can be put together, and of an environmentally less sensitive area. for the French the problem is more that care must be taken not to disturb finely balanced man-made drainage systems rather than major concern for landscape, flora and fauna. There is a danger, however, that while the need for care of the environment in Kent is acknowledged it is less obvious in France, and too little care may be taken as a result. The development is seen as more of a technical than an aesthetic problem in France. Probably too, environmental issues are of lower priority in public opinion generally in France than in Britain.

However, Environmental Impact statements have been obligatory in France since 1976; in Britain there is as yet no statutory requirement to provide one.

In France the tunnel exit is dealt with separately from most other developments. The two problems which seem to have caused most concern during the public utility enquiries are the disposal of spoil - now estimated at 4.5 million cubic metres - by hydraulic means, and which gives cause for concern about possible deposits of salt mud and effects on retaining dykes, and the effects on the water system. The agricultural community is also concerned about the risk of flooding on neighbouring farmland, and has successfully demanded managerial and financial guarantees from Euro Tunnel. These have been underwritten by the State.

2(c) Wider strategic implications of the Tunnel

Discussion in the group at this point centred very much on transport and probably considered aspects discussed in much greater depth of the Transport Workshop.

The importance of an improved total infrastructure was stressed. In France, for example, it was found that the TGV released considerable latent demand for fast, efficient travel, and so increased general mobility. It was stressed though that the combination of factors that make such a venture a success are not quite clear and may not, of course, be repeated for the tunnel. But it was felt that this experience made the French see the tunnel as only part of their improving transport network - from the Channel to Spain, the west coast and to other European countries.

From the French experience it was felt that, contrary to popular belief in Britain, the Channel Tunnel could possibly do _more_ for regions of Britain away from Kent than for Kent itself - _if_ it is allied to an improved network. There is a great deal of misinformation and speculation about this.

It was generally agreed, therefore, that the shuttle train is not the important factor; the really important development is the main railway with through passenger and freight trains. Nevertheless, the shuttle is the driving force for the profitability of the tunnel, thus creating a serious contradiction.

There was a difference of opinion amongst the British members of the workshops over whether there really is forward and strategic planning going on at British Rail. There is certainly disquiet in northern England and Scotland that the Tunnel will act as a magnet to the South-east and further disadvantage these other regions. There seems to be no doubt that there _is_ strategic thinking on both rail and road in France.

The British group were asked if there is a 'strategic plan for Britain', or perhaps for Kent, with respect to the Tunnel. There is no overall strategy for Britain, but Kent County Council had insisted on the setting up of an Impact Study Group which consists of representatives from central, county and local governments, Eurotunnel, and British Rail. The objective is for the Group not only to investigate likely impacts but also then to work out together plans for capitalising on the advantages of the tunnel and also, perhaps more importantly, to take positive steps to alleviate the disadvantages.

It is difficult for governments, British Rail or SNCF to anticipate exactly what developments will stem from the Channel Tunnel and therefore invest ahead. What will be necessary is co-operation and flexibility in order to seize opportunities as they are created and support their development.

2(d) The organisation of major construction sites

After participants had visited the sites for the tunnel workings and operations the problems for construction on the
British side, and why the tunnel and its facilities cause
planning and environmental problems in Kent. In view of
this it was decided to consider the organisation and site
control of major projects. Again, there are marked
differences between the French and British approaches.

In France there is a special procedure for all large sites
which has been developed by EDF from their experience of
building nuclear power stations. The procedure is co-
ordinated by a state official and is designed to solve three
problems:

- the accommodation of construction and other workers
- training in the skills required for construction
- retraining and dispersal on completion of the project.

This has emerged as a common law process in the 1970s to
ensure close collaboration between local authorities, firms,
educational bodies and the State. The Delegation de
l'amenagement du Territoire (DATAR) has to approve all
procedures and co-ordinate ministerial intervention with
the Prime Minister as arbitrator. Manpower is provided by
DATAR and finance by the amenagement du territoire budget,
the tax revenues of the communes and special, deferred loans.
In all this special attention is paid both to the aftermath
of development and to ensuring that the area concerned gets
a fair share of jobs and contracts, provided that this does
not impose excessive on-site costs.

This procedure will be followed although the tunnel will
create some problems mainly because it is a private sector
development, but also because of other aspects such as that
Eurotunnel have agreed to take 70% of its labour force from
Nord Pas de Calais.

Britain, on the other hand, has no pre-determined procedure for the co-ordination of the public sector involvement (such as government agencies, police, fire services) in a private development. In this case Kent County Council is, in fact, co-ordinating but this is a voluntary service and confers no powers on the County Council. For example, there is no way that Eurotunnel can be forced to take labour from the local region, KCC can only encourage Eurotunnel to recruit locally by trying to make it easier for them, such as by joining with Eurotunnel to provide a local training centre for workers. Eurotunnel can, however, recruit from anywhere including other countries if it so wishes.

The South-east region of England does not therefore view the employment opportunities so favourably as the Nord Pas de Calais. There is a mismatch of skills between the labour force in the South-east and those required. A positive attitude toward the employment opportunities offered by the tunnel is somewhat over-shadowed by long-term prospects of consequent unemployment in the ports. Nord Pas de Calais can view much more positively the creation of 4000 jobs per year during the construction phase.

3.   Conclusion

Members of the workshop closed their very interesting discussions by noting how different in many ways were the opportunities, objectives, organisations and problems of France and Britain, in attempting to achieve the same end result. But that also how closely those involved had been able to work together, particularly those from Kent and Nord Pas de Calais at either end of the tunnel, and how successful they had been in co-ordinating and integrating their activities.

IV. TRANSPORT AND EUROPEAN INTEGRATION

a) Position Paper

TRANSPORT INVESTMENT AND EUROPEAN INTEGRATION

By Roger Vickerman

The lack of any fixed transport link between Britain and
continental Europe has both an economic and a cultural significance. This paper aims to explore the extent of that significance, to pose a number of key questions and to draw some
preliminary conclusions about the validity of the emphasis on
infrastructural investment in the Common Transport Policy as an
aid to integration.

In strict economic terms the significance is felt in the higher
transport costs experienced by both importers and exporters.
This involves the costs charged by ferries, the higher costs due
to any extra handling, and the higher costs arising from the
incentive to use higher cost modes such as road transport because
of the cost penalty imposed on railways due to the need for mode
changes (e.g. for a typical haul of 600km on the continent rail
would have a cost advantage, for such a haul between a British
and a continental location road obtains an artificial advantage
because of the bias to road of ro-ro ferries). Costs are not
just direct financial costs, but also involve the extra time
costs involved in mode-changes. These costs are significant
both for British exports and imports and for the European economy
as a whole since the imposition of additional transport costs
works just like a tariff barrier and reduces the volume of trade
in total with effects on economic growth.

In addition to these more observable and measurable costs there
is undoubtedly a significant subjective cultural dimension which
also causes trade patterns to differ from the ideal. Partly
this is an effect which any frontier would have. There is
evidence, for example, that the Franco-German frontier poses
an enormous barrier to trade, perhaps being equivalent to as

56

much as the transport costs of 600km haul (Peschel, 1981). Partly it arises from Britain's traditional trade pattern having been much more strongly related to overseas than European countries (both North America and the rest of the Commonwealth). And partly it arises from pure cultural differences which lead to British consumption patterns and relative prices being substantially different from those in other countries.

Against this background we need to answer the question as to how far the completion of a fixed link, with its ensuing reduction in at least the perceived costs of crossing the Channel, will lead to a fundamental change in these relationships. We need to assess these effects on two levels, the differential effects on individual regions (especially those in close proximity to the link) and the aggregate effect on the European economy as a whole.

One of the principal arguments which we shall advance is that it is impossible to assess the Channel Fixed Link on its own. We have already shown elsewhere (Vickerman, 1987a, 1987b) that these direct effects are likely to be relatively small in net terms, but two further developments are relevant. One is the development of Tunnel-related infrastructure which raises the impact of the Tunnel - the best example of this would be the completion of high-speed rail links between the Tunnel and Paris/Brussels. The other is the interest in other parts of Europe in major infrastructure projects which if undertaken in the absence of the Channel Link would cause an adverse movement in relative transport costs and accessibility to British locations. Example of these include other major fixed links such as the Great Belt crossing in Denmark, the Oresund crossing between Denmark and Sweden, new Alpine base tunnels, and the creation of a European high-speed rail network which crosses existing frontiers, of which the French Lignes à Grande Vitesse and German Neubaustrecke are initial components. The possibility of a Paris-Brussels-Amsterdam/Koln link is itself closely related to the Channel Tunnel project.

Such an argument is important in so far as it requires us to consider the definition of the counter-factual situation

carefully. There is a particularly important additional dimension
to this when there is the possibility of competition for available
private investment funds, given the increasing interest of European
governments in reducing the role of the public sector. So far the
French TGV (Sud-Est and Atlantique) and German NBS and ABS have
been largely (though not exclusively in the French case) financed
by their respective governments. The Channel Link is to be totally privately financed (as are other possible new infrastructure
investments in the U.K., c.f. the discussion in Vickerman, 1987c),
the French government has indicated that it may require the TGB-
Nord to be privately financed. The international group examining
possible international extensions of the TGV-Nord involving
Belgium, the Netherlands and West Germany has also started to
grapple with the problem of who pays for international infrastructure - in proportion to distance covered within the
country's territory or in proportion to likely benefits derived.
This is also of interest to Britain in that British benefits
from the Tunnel clearly are related to infrastructure provision
on the continent. The British position, however, has been
stated by the David Mitchell the Minister for Public Transport,
as being one of provider pays. This could result, for example,
in Belgium paying the largest contribution to the new network in
order to speed traffic through its territory and reduce the
economic benefit it receives.

In the remainder of this paper we post four basic questions:

(i)   how important are transport infrastructure
        improvements for the economic development
        of Europe in general?

(ii)   how are the benefits and costs of a particular
        infrastructure improvement distributed between
        regions, especially between those regions
        directly involved in the infrastructure and
        those more distant?

(iii)   how significant are non-pecuniary factors in
        determining the total effect - i.e. is there
        a 'psychological' integration effect to be
        considered?

(iv)   how should international infrastructure (or

national infrastructures which benefit those
outside the country) be paid for?

1. Aggregate Significance of Transport Infrastructure

The basic argument in favour of infrastructure in the aggregate
is that better infrastructure reduces transport costs. Transport is a major contributor to the economy, both in its contribution to GDP and in the labour employed, but these are both
essentially input rather than output measures and from the point
of view of aggregate economic growth it may be more efficient to
redice these measures of the size of the transport sector if its
real output can be increased. It is the productivity of the
transport sector which is critical rather than its absolute size.

Transport infrastructure is itself extremely difficult to measure
(c.f. Biehl, 1986) and the critical factor is likely to be not
its aggregate size or capacity but its connectivity, accessibility to it and the presence or absence of bottlenecks (Blum, 1982
and c.f. discussion in Vickerman, 1987d).

This confusion about the true role of transport infrastructure
can be noted in two recent statements by EEC officials. In a
lecture in March 1986, Pierre Mathijsen, Director General for
Regional Policy, summed up the new directions in regional policy:

> 'in one word: concentration. Since the means at the
> disposal of the Community are very modest indeed,
> efficient use can only be insured by concentrating
> their allocation in a limited area at one time;
> dispersion means limited effect and therefore
> inefficiency. On the other hand, more attention
> should be given to *productive investments rather*
> *than infrastructure'*. (Mathijsen, 1986, emphasis added)

On the other hand during a visit to tunnel sites in Kent and
Nord-Pas de Calais in February 1987, Stanley Clinton Davis,
Commissioner for Transport and the Environment is quoted as
saying:

> 'I welcome major transport infrastructure projects
> because of the stimulus which they provide to jobs

and the economy as well as the role they play
in the development of a modern and efficient
transport network within the Community. The
crucial part which such projects can play in
economic regeneration is as important at a
national level as it is at Community level -
although not all governments deem to realise this'.

The spillover effects of investment particularly have led to a
concentration of EEC interest in the consistent development of
the transport network (Commission of the European Communities,
1979) and especially in the consistency of any cost benefit
assessment (Commission of the European Communities, 1973).
The aggregate effect of any one piece of infrastructure is,how-
ever, highly marginal to the overall level of activity, except
when certain conditions hold. First, is the case of the bottle-
neck; improvement of certain pieces of infrastructure can have
major impacts on a very wide area such as the Severn Bridge,
the St. Gotthard Tunnel and, on a slightly larger scale, the
M25 around London. Secondly, there is the case where the
improvement generates considerable new economic activity in a
region which has major spillover effects on other regions. An
example of this might be pipeline construction enabling exploration of oil or gas fields such as in the Groningen region
of the Netherlands. Thirdly, there is the case where a number of
pieces of infrastructure are developed consistently thus changing accessibility in a number of regions simultaneously and
hence having a major effect on the economic potential of an
area (c.f. Keeble et al, 1982). This effect is accentuated by
any associated change in technology. There is evidence, for
example, of major changes consequent on the building of high-
speed rail lines and even just from railway electrication. Here
the critical factor may be as much subjective and/or quality
related as deriving from quantifiable economic factors such as
cost or time.

2.   Distribution of Benefits

The distribution of benefits from a particular piece of infrastructure investment can be considered both between regions and

within regions. The inter-regional effect is the core of the basic economic argument over integration since it is saying that a reduction in transport costs is a reduction of a cost barrier to trade which will open up trade possibilities. Transport costs thus serve just like tariffs in hindering integration.

One of the major problems with assessing the effects of transport cost reductions, however, is that existing patterns of industrial location, regional production and consumption, reflect an optimising adjustment to the existing transport cost pattern. Whilst at an aggregate level lower transport costs would seem simply to increase trade according to the usual tenets of comparative advantage, at a disaggregate, industry by industry, level, the adjustment may be much more complex. This depends partly on the significance of transport costs in each industry's output and partly on the scope for substituition between transport and other inputs to the production process.

Initial attempts to try and assess the effects of the Channel Tunnel on adjacent regions suggest that the scope for much industrial relocation is extremely limited, especially in manufacturing industry. Two exceptions to this would be cases of highly mobile new investment, typically inward foreign investments, and of industries which face severe constraints on development in existing locations, such as high technology industry in the original high-tech corridors along the M4 and M11 in Britain. However, there are two further areas of potential Tunnel-related growth which are worthy of note: distribution and tourism.

Distribution is a very diverse sector and whilst we typically think of its major components as the wholesale and retail distribution of finished goods it also features in a significant way within the production process. It may well be that significant changes in both wholesale distribution and retailing, taking advantage of changing transport provision, will take place. However, of much greater potential significance for integration would be changes in production patterns to take advantage of such changes, using the transport system to locate cheaper sources of materials or to lead to greater specialisation in certain locations; in effect substituting transport for other factors of production. Perhaps one industry to take early advantage of this could be the motor industry since it is already organised on an international basis and firms such as Ford, General Motors and Peugeot are already using plants in different countries to specialise in the production of specific parts or specific models.

Tourism has been a major growth industry in much of Europe, both internally and for visitors from outside Europe. It is also seen by many as a major beneficiary of improved communications. It is not, however, totally clear how price sensitive tourism is in the aggregate. Changing relative access prices may change the distribution of tourist destinations without necessarily increasing the aggregate amount of tourism. More general economic indicators of wealth and unemployment are more important as determinants of total tourist traffic. As with manufacturing industry therefore the critical factors will be, by how much do access costs for tourists change and is there a major change in the perception of areas as potential tourist locations. Whilst the direct benefits for the regions adjacent to the Tunnel may be somewhat limited as a result of this, it is possible to expect a bigger change in the type of transport used for tourist traffic between the U.K. and continental Europe in general which could lead to an indirect increase in tourist traffic in Kent and Nord-Pas de Calais from stopover traffic.

Within the regions concerned there may also be important changes. The Tunnel would seem to draw the benefits of locating in the region away from the immediate port and coastal areas and towards locations perhaps 100-150km inland where the benefits of access to other regions and the effect of other infrastructures become more significant. Hence, whether or not there is a major change in the economic fortunes of the Tunnel-mouth regions there is certainly likely to be a change in the distribution of welfare within those regions. Such a change may in the last analysis only involve the reinforcement of a general trend away from relatively peripheral and inaccessible coastal areas which is strongly evident in Britain anyway, but the existence of the

Tunnel is undoubtedly unlikely to reverse this trend and may well accelerate it, leading to strong claims for compensation and the perception that the tunnel has damaged local interests.

3. Perceived and Subjective Effects

We have argued above on several occasions that the most important integrational effects could be due to a change in the perceived accessibility of locations rather than to any real change. We have also suggested that conversely the Tunnel could receive the blame for unfavourable trends in the local economy. It has become almost ritualistic now to blame the EEC for many of the ills of the british economy, and whilst there is some truth in the view that poor performance in certain sectors can be associated with the increased competition and penetration of domestic markets resulting from Community membership there is other evidence e.g. on relative performance in third markets which suggests that this decline is independent of membership. Many of the fears expressed over the Tunnel, reflect this view that imports will benefit much more than exports and British tourists to the Continent much more than continental tourists to Britain. How far this is ultimately true is likely to depend as much on marketing of Tunnel-related opportunities as on anything else.

As well as these less rational aspects, there is also a range of rational economic factors which are nevertheless difficult to measure objectively. We do not know how much to value the cost-penalty implied by seasickness, nor indeed how much the equivalent Tunnel problem of claustrophobia is worth. It will be interesting to see how perceptions of the relative safety of ferries and Tunnel change as a result of the Zeebrugge ferry tragedy, or whether this simply raises the perceived isolation of Britain whatever means of transport are used. We do know that comfort and reliability are valued highly by customers and we know that for both passenger and freight traffic time penalties imposed by a change of transport mode are valued much more highly than pure journey times. This reinforces the suggested view that through rail traffic has most to gain from the Tunnel system, the fact that it will be possible to travel from a British location to a continental location without a

change of mode (even if the actual locations linked in this way
are in practice very limited) could involve the most significant
change in perceived accessibility of all.

4.   Paying for Infrastructure

This is not the place for a full analysis of the question of
private versus public finance for such projects (see Vickerman,
1987c, for a fuller discussion). We have, however, already
raised the difficult issue of the distribution of costs relative
to the distribution of benefits and this could be a key issue in
both the actual and perceived effects on integration. It raises
essentially the argument as to whether the benefits from infrastructure are so strong and so widespread that all infrastructure
should be treated as a public good, financed from general tax-
ation revenue and available free to all. Since the usage of
large parts of infrastructure is effectively both competitive
(because of congestion) and price excludable (especially for
links where there is no adequate free alternative) this argument
clearly has limittions. It is clear that infrastructure can
be privately financed implying that there is an adequate
financial return which in turn implies no market failure, but
this does not necessarily mean that it should be so financed.

Two questions are suggested here. One is the fear of mono-
polistic exploitation of key links. Here in the Tunnel case the
argument can be used both ways in that Eurotunnel see themselves
as introducing competition onto a route that is now under the
control of a thinly disguised cartel, but Flexilink has argued
that the long term economics of competition between Tunnel and
ferries is inherently unfair (in the Tunnel's favour) leading to
a Tunnel monopoly which is stronger than any the ferries could
devise because of the costs of entry. This argument is a much
wider one, however, since transport by all modes across Europe
is a highly regulated industry and one characterised by monopolies and cartel, frequently state-owned. The major source of
difficulty in the search for an acceptable Common Transport
Policy within the European Communities has arisen from this deep
and complex state involvement in airlines, railways and the

regulation of inland waterways and road haulage. Would a private sector monopolist be less efficient than the many state-owned ones in this sector?

The second question is that even where social benefit-cost analyses are undertaken of transport projects they tend to concentrate on the direct secondary effects on local employment and the environment and not on the deeper questions of integration. Whilst in the search for a Common Transport Policy the Community, often in frustration because of the lack of progress on the key issue of creating the Common Market in transport, has turned to the question of defining European networks, it has never got much further than identifying bottlenecks and specifying common methods of appraisal. The key question of the role of transport in achieving the Common Market in general terms, although identified in the Rome Treaty and in many of the subsequent attempts to put a Policy into operation, has never been adequately answered. Perhaps in the last analysis that is why we are still faced with so many dilemmas in considering the Channel Link and why prejudice and suspicion have so often dominated the economic case.

References

Biehl, D. 1986, *The Contribution of Infrastructure to Regional Development,* Final Report of Infrastructure Study Group, Commission of the European Communities, Luxembourg.

Blum, U. 1982, Effects of transportation investments on regional growth: a theoretical and empirical investigation. *Papers and Proceedings of the Regional Science Association,* 49, 169-84.

Commission of the European Communities, 1973, *Coordination of Investments in Transport Infrastructures,* Studies, Transport Series No.3, Brussels.

Commission of the European Communities, 1979, *A Transport Network for Europe: Outline of a Policy,* Bulletin, Supplement 8/79.

Keeble, D., Owens, P.L. and Thompson, C. 1982, Regional accessibility and regional potential in the European Community,

Regional Studies, 16, 419-32.

Mathijsen, P. 1986, New Directions in European Regional Policies, Discussion Papers in European and International Social Science Research, University of Reading, No.12, May.

Peschel, K. 1981, On the impact of geographic distance on the interregional patterns of production and trade, *Environment and Planning A,* 13, 605-22.

Vickerman, R.W. 1987a, The Channel Tunnel and regional development: a critique of infrastructure led growth, *Project Appraisal,* 2, March.

Vickerman, R.W. 1987b, The Channel Tunnel: consequences for regional growth and development, *Regional Studies,* 21, June.

Vickerman, R.W. 1987c, Private Finance of Public Projects, mimeo, University of Kent at Canterbury, January.

Vickerman, R.W. 1987d, The Impact of Transport Infrastructure Investment on Spatial Structure, A Report to the Institut de Récherches Economiques et Régionales, Université de Neuchâtel, Switzerland, March.

Three main themes were identified from the paper and a Workshop
session was devoted to each of these:

(i)   the spatial scale of the effects of the Tunnel,
        whether it will be at a purely local-regional level,
        at a wider regional-national level or have its great-
        est significance at a European level;

(ii)   how far it is possible to quantify these effects on
        the basis of objective criteria or whether the key
        impact of the Tunnel will be the result of changes in
        attitudes, both to perceived reductions in distances
        and to changing views of peripherality and national-
        ism;

(iii)   the financial questions posed by the Tunnel, not
        simply the effects of the choice of private rather
        than public finance (more fully discussed in another
        Workshop), but the implications for the public sector
        of the project in terms of its externalities and the
        demands for compensation.

The discussion on these themes had to remain speculative since it
was felt that there is insufficient evidence to date to make a
start on a full assessment of these questions, but in many
respects these were felt to be the key aspects to the success of
the Tunnel project. If it fails to have an impact outside its
immediate locality and fails to change people's attitudes or
perceptions then most of the political rhetoric used to justify
the project will have been in error. However, the discussion
benefitted by having present officials from the planning authorities both in Kent and in Nord-Pas de Calais who reminded participants of the immediate political and administrative realities.

1.   The Spatial Scale

The main thrust of the argument in this Session was that the
Tunnel is, of itself, only a tunnel, what makes it significant is
the regional transport network which connects to the Tunnel and,

possibly more importantly, the wider international network of which it becomes a part. The extent of its impact will depend on the quality of these networks and the ensuing impacts on both the relocation of enterprises and people and on their interaction through trade flows etc.

Three levels of impact were identified, the local-regional, the regional-national and the Europe-wide. Much of the research to date has concentrated on the local-regional level, not surprisingly since this has largely been the concern of the local-regional planning authorities who have to plan for the immediate surroundings of the Tunnel. Clearly there will be a major impact at this level, but its extent and significance could well depend on what happens outside the two neighbouring regions. A particular issue of concern here is whether the regions next to the Tunnel simply suffer from a 'corridor effect' with any greater ease of crossing the Channel reducing the need either to locate, or make stops, in Kent or Nord-Pas de Calais.

At a regional-national level we are looking at a slightly wider impact including the need to make more direct comparisons between Kent and Nord-Pas de Calais and the way these regions relate to their own national economies. The emphasis here was on the differences between the regions in both economic and environmental character. This raises important questions as to the extent to which any real economic integration can take place; if they had mutually beneficial influences on each other, would these not have been developed already given the existing close links provided by ferry routes? A further clue is given by the regions' positions within the U.K. and France respectively. Both are peripheral regions, but Kent is both integrated into the dominant influence of London in South East England (although East Kent has performed less well economically than most of the South East Region and has pockets of very high unemployment) and is the closest part of Britain to continental Europe. Nord-Pas de Calais, on the other hand, is much more isolated in France, outside the influence of Paris and, despite its more central location in Europe, suffers from its position on a frontier (with Belgium) which reduces its natural hinterland. As a measure of

68

economic situation, Kent has unemployment rates at about the
national average, 12.1% (though a little above the South East
Region average of 9.6%), whilst Nord-Pas de Calais has an unemployment rate of 14% well above the French national average
(although the rates in East Kent and in Nord-Pas de Calais are not
significantly different from one another).

At the wider European level it is much more important to think in
terms of the way the Tunnel fits into the pattern of communications between all the countries of North-West Europe, including
Belgium, the Netherlands and North Rhine Westphalia in the Federal
Republic of Germany. Following a long period of relative in-
action there is renewed interest in trying to make the European
Community's Transport Policy work towards integrating national
economies across national frontiers. Proposals affecting rail
developments are particularly important here. If the existing
national frontiers are ignored it is possible to conceive of a
major region lying centrally within Europe which needs improved
communications both within it and between it and other regions
in order to promote integration and economic development. One
factor which is thought to be particularly important in this is
the way in which Spanish and Portuguese accession to the Community
have altered both its geographical centre and the directions of
some of the main potential axes of communication. Whilst the
creation of such a new Euro-region may benefit Nord-Pas de Calais
and certain Belgian regions such as Hainaut and West Flanders more
than Kent, because of their ability to develop crossroads between
these axes, Kent may be seen to have certain advantages which
would include this part of South East England into a 'Channel-
side' Region. The process of development would depend both on
consumption and production effects. The consumption effects would
derive from new trading patterns made possible by improved transport lowering costs and enlarging markets. The production effects
would derive from the way in which new and improved transport
facilities stimulate changes in production processes and make
more integrated production possible between different countries,
thus using transport more effectively as a factor of production.

In the discussion one point stood out, regardless of the spatial

scale, the critical importance of distinguishing the through rail
services from the Tunnel-shuttle service. The latterwas argued
to be significant only for its competition with existing ferry
services, it would be the former which wouyld forge the new links
and make possible greater integration between the neighbouring
regions and have wider European significance, whilst potentially
reducing the environmental impact of road traffic within the
regions. However, it was also recognised that the extent of this
impact depended both on the respective railways, British Rail and
SNCF, being able to rise to the challenge of finding and exploiting new markets and on their ability to secure a financial settle-
ment with Eurotunnel as Tunnel operators which would allow them
to be competitive. In this light the development of rail networks
beyond the Tunnel would become critical, especially the TGV-Nord
with its proposed links to Paris and Brussels and beyond towards
Amsterdam and Koln.

A further issue, on which there was much less agreement, was the
significance of the Tunnel project in the moves towards completion
of the internal market in the Community. Whilst it was accepted
that much of the existing delays to Franco-British freight traffic
occurred through frontier controls rather than transport delay,
which the Tunnel of itself would not change, but which could
severely limit the integrational effects of the Tunnel, there was
not widespread belief in the view that completion of the Tunnel
would give impetus to removal of such controls.

2.   The Subjective Impact

The outcome of much of the research into the likely impact of the
Tunnel to date is that it is extremely difficult to identify
significant changes in the hard economic factors affecting
location and trade flows. However, the Tunnel's primary importance has been suggested to be in the way it will affect people's
attitudes. If people believe they are a long way apart, even
when they are not, they will not trade with each other to any
great extent as we have seen within continental Europe. On the
other hand, if people can be made to believe that the Channel no
longer poses the obstacle to economic interaction it has hitherto
been thought to, they may change their behaviour much more

significantly than any analysis based on extrapolation of current trends may suggest. This Session was largely devoted to this theme and became understandably much more speculative than the previous one.

The dimensions of this change are similar to that developed in the previous section and can be considered at the same three spatial levels. In Nord-Pas de Calais in particular it could be felt that the Tunnel had a special significance at the local level in raising the profile of a generally depressed region and redressing some of the traditional imbalance between it and other French regions. However, it was also pointed out that the population of Nord-Pas de Calais had been seen to be fairly resistant to change and was perhaps the least mobile of all French regional populations. Hence, even at the fairly simple level of evening up the present imbalance in excursion traffic across the Channel the existence of the Tunnel was thought unlikely to provoke a greater interest amongst local French residents. Nevertheless it was also felt that there was evidence of considerable nationalist-ic pride in building things on the French side which would awaken general interest in the project and that this might be reflected in increased regional dynamism.

In Kent too there was some evidence of changing attitudes towards the Tunnel with a recognition that it could be used as a means of imbuing this same regional dynamism, particularly from towns such as Ashford and Folkestone which would have the most direct relationships with the Tunnel.

At the wider European level the move towards international trans-port infrastructure projects, the renewed interest in the role of transport policy and the current concern with moves towards completion of the internal market could all be seen as aspects of a renewed interest in the more fundamental aspects of inte-gration and how to bring it about. However, at an individual level there was still a question as to whether stressing the European integration aspects of the Tunnel could ultimately be counter-productive to its success given the general mistrust of the EEC as an institution, particularly in the U.K.

3. Financial Questions

The major financial question addressed was concerned with who actually pays in the final analysis? This requires us to look carefully at the distribution of benefits and costs from the project. Again the same three spatial dimensions are relevant since we can look at the regional incidence within and between the countries and the intra-regional effects within each country. This recognises that within regions the areas to benefit may be at some distance from the Tunnel itself whereas the main direct costs fall on the port communities and those affected by construction and operation of the terminals. However, there may be public finance implications of this involving inter-governmental grants and the distribution of associated infrastructures may also involve important public finance flows.

The emphasis on the importance of associated networks and infrastructures gives a further clue to this problem. If to be a success the Tunnel system requires such networks then much of the burden of creating this success will fall on the providers of such networks, basically the public sector. In one sense, therefore, these associated public works could be viewed as a substitute for a direct financial guarantee to the project from the Governments. The dividing line between being a subsidy to the project as a whole and compensation for any negative externalities from the project could be a very thin one. This point was looked at in the context of some of the main transport infrastructure improvements proposed in the two regions.

In the final analysis the critical factor will be the extent to which the external benefits exceed the external costs, that depends mainly on the broader economic consequences although the redistributional effects of traffic from road to rail could also be significant here, but even if this balance is hugely positive there could be awkward redistributional aspects which have not yet been fully addressed.

Some Conclusions

The main conclusion of the workshop was one of inconclusion, that it remained extremely difficult to assess the wider impacts

72

of the Tunnel, precisely because they were potentially very wide indeed. However, it was agreed that it was important to distinguish between the various spatial dimensions in order to categorise the impacts, that the various networks linking the Tunnel into a wider economic environment would be significant, that much of the effect could result from changes in attitude rather than changes in hard economic variables and that the barrier between public and private sectors and public and private finance remained extremely difficult to identify.

V.   AFTERWORD

Although the Colloquium ended on this slightly inconclusive note,
opinion was clearly on the side of trying to fill some of the
many gaps in our knowledge of the likely effects of the Tunnel.
So, as well as encouraging the idea of publishing these papers,
it was hoped that subsequent developments would indeed be proper-
ly studied and monitored. Some collaborative proposals for doing
this are already under consideration between Kent and Lille, but
for the many others who will be interested, including members of
UACES for whom its European potential will be a major concern,
it may be helpful to indicate a few more additional sources and
readings beyond those specifically cited in the position papers.
These do not cover ordinary press coverage but are limited to
some of the more substantial books, reports and articles in both
languages which have come out in the last few months, some of
which have come out since the Colloquium was held. No doubt
there will be more such works in the months to come, especially
now the Ratification is complete and scholars, administrators
and the general public will have to face up to the reality of the
Tunnel, but this sample gives some indication of the materials
available at the time of writing.

Finally, we are most grateful to all those who, by helping to
fund and organize the Colloquium and this publication, gave us
the chance to start thinking about the context of the Tunnel.
We hope that our tentative attempts to go beyond the merely
polemical and to come to grips with some of the questions which
the approach of the Tunnel seems to raise and the ways that
these may best be answered will help to guide and stimulate
others to explore matters further. We hope that, as the
construction of the Tunnel gets underway, there will be a
serious effort to understand and exploit its potential, not
merely in Britain and France, but elsewhere in Europe as well.

> August 1987: Clive H. Church, for
> the University of Kent, the Institut
> d'études politiques de Paris, and
> UACES.

READING LIST

M. Anderson   'The Channel Tunnel. A Cause for Optimism',
The Planner 72 (1986) 15-17.

M. Bonavia   *The Channel Tunnel Story* (Newton Abbott,
David and Charles, 1987).

P. Bruyelle   'Le tunnel sous la manche et l'Aménagement
Régional dans la France du Nord',
Annales de Géographie 534 (Mars-avril,
1987).

Channel Tunnel Act 1987 cap. 53 (London, HMSO, 1987).

L'Européenne   Special number on 'Le dynamique européenne
des transports' July-August 1986.

Eurotunnel   *The Channel Tunnel. The vital facts* (London,
January 1987).

P. Gallois   *Les Grandes étapes du lien fixe Transmanche.
Etude bibliographique* (Wissant, Syndicat
d'Initiative, 1986).

E. Grzelal   'The Channel Tunnel' *European Trends
1986/4* (London, Economist Intelligence Unit).

N. Henderson   *Channels and Tunnels. Reflections on
Britain and Abroad* (Weidenfeld & Nicholson)
Feb. 1987.

M. Hamer & B. Jones   'Tunnel will link to three Continents',
New Scientist (9 April 1987)

S. Harrison   *The Channel* (London, Macmillan, 1986)

House of Lords   *Special Report from the Select Committee
on the Channel Tunnel Bill* 138 (London,
HMSO, 6 May 1987).
*Select Committee on the Channel Tunnel Bill;
Minutes of Evidence* 138 1-111 (London,
HMSO, 6 May 1987).

Joint Consultative   *The Kent Impact Study. A preliminary
Committee   assessment.* (London, Department of
Transport, 1986).

Joint Consultative Committee	*Kent Impact Study. The Channel Tunnel A Strategy for Kent* (London, Department of Transport, August 1987).
B. Jones et al. (eds)	*The Tunnel. The Channel and Beyond* (Chichester, Ellis Horwood, 1987)
Kent County Council	*The Channel Tunnel and the Future for Kent* (London, Department of Transport, 1986) *Structure Plan. Second Revision. Draft Written Statement* (1987).
Lille Actualités	Special Number 'Tunnel: le Chantier du Siècle' (March 1986).
D. Mitchell	'The Tunnel' *Oast to Coast* Summer 1986.
J-P. Navailles	*Le Tunnel sous la Manche. Deux siècles pour sauter le pas (1802-1987).* (Seyssel, Editions du Champ Vallon/P.U.F., April 1987).
Région Nord-Pas de Calais	*Le Tunnel. Bulletin spécial d'Information Le Tunnel sous la Manche. Quel intérêt pour les enterprises du région? Forum du 10 octobre 1986.* (Lille, 1987). 'La Liaison-Fixe Transmanche: Hier, Aujourd'hui, Demain' *Nord-Pas de Calais Informations* 2.86/06. *Lien Fixe Transmanche. Eléments pour un plan de développement de la région Nord-Pas de Calais* (Lille, January 1986). *Plan Transmanche. Protocole d'Accord Etat-Région Nord-Pas de Calais* (Lille, March 1986).
M. Simmonds	'The Channel Tunnel. For Better or for Worse' *The Planner* (March 1987)
A. Stevens & P. Holmes	*The Framework of Government-Industry relations in France* (University of Sussex Working Paper, 1986).
R.W. Vickerman & C.H. Church	'The Impact of Frontiers. A British Perspective', forthcoming in *Les Cahiers du Craps* 4, (Université de Lille II).